The Trip of a Drip

Other books by Vicki Cobb

Chemically Active! Experiments You Can Do at Home
Making Sense of Money
The Monsters Who Died: A Mystery about Dinosaurs
Science Experiments You Can Eat
Magic . . . Naturally: Science Entertainments and Amusements
More Science Experiments You Can Eat
Bet You Can! Science Possibilities to Fool You (with Kathy Darling)
Bet You Can't! Science Possibilities to Fool You (with Kathy Darling)
The Secret Life of Cosmetics
The Secret Life of School Supplies
How to Really Fool Yourself: Illusions for All Your Senses
The Secret Life of Hardware: A Science Experiment Book
Lots of Rot
Fuzz Does It!
Gobs of Goo
How the Doctor Knows You're Fine
Brave in the Attempt: The Special Olympics Experience
Supersuits
The Scoop on Ice Cream
Sneakers Meet Your Feet
More Power to You

The Trip of a Drip

by Vicki Cobb
illustrated by Elliot Kreloff

Little, Brown and Company
Boston   Toronto

Text copyright © 1986 by Vicki Cobb
Illustrations copyright © 1986 by Elliot Kreloff
All rights reserved. No part of this book may be reproduced in any form or by any electronic or mechanical means including information storage and retrieval systems without permission in writing from the publisher, except by a reviewer who may quote brief passages in a review.

Library of Congress Cataloging-in-Publication Data
Cobb, Vicki.
   The trip of a drip.

   (How the world works series)
   Summary: Trace the journey water makes from, to, and through our homes.
   1. Municipal water supply — Juvenile literature.
2. Drinking water — Purification — Juvenile literature.
[1. Drinking water. 2. Water] I. Title. II. Series.
TD348.C63 1986   628.1   85-23960
ISBN 0-316-14900-4

*Published simultaneously in Canada
by Little, Brown & Company (Canada) Limited*

Printed in the United States of America

This series is dedicated to Louis Sarlin,
   the teacher who gave me the best year of my childhood
   and the key to my place in the world.

Contents

1. From "On" to Gone —————— 1
2. Water for a Country Home ——— 5
3. Water for Cities and Towns ——13
4. From Sewer to Stream ————24
5. Traveling in Circles ——————34
   Some Tricks
       with Your Drips ——————41

The author gratefully acknowledges the help of the following: Lisa E. Przybylowicz, of the State of New York Department of Health; James Neary and Marianne Vogel for their tour of a water treatment facility; Assistant Commissioner Robert Wasp; John Karell, Jr., P.E.; Louise Carosi and Spencer Conklin of the Westchester County Department of Health; and Malcolm Beal, Jr., well-driller, of P. F. Beal & Sons, Inc.

The Trip of a Drip

1.
From "On" to Gone

Want a drink of water? Want to wash your hands? No problem. Just go to the sink. Turn the water faucets to "On." Out comes pure, clean water. Clean enough to drink. You get plenty of it, too. More than you need. Extra water is wasted down the drain. You usually don't think twice about it. It's gone.

How simple and easy it is to get water when you want it! You don't have to lug heavy buckets back from a well. You don't have to walk to a river to take a bath. You don't have to light a fire under a tub to get hot water. Clean hot and cold water is at your fingertips.

You don't have to worry about getting rid of water you don't want, either. You don't have to carry away buckets of dirty water to dump into the ground. On the street you don't have to step over smelly sewers that carry away waste and dirt. Flush the toilet. Pull the plug. Good-bye, waste. Talk about clean living!

Water is something you *must* have to live. A day without water and you will have a painful thirst. Two days and you will feel weak. You cannot think clearly. You see strange sights that aren't there. Three days without water and you could be near death. You need to take in about two quarts of water every day. Half comes from liquids you drink, like milk, soda, and, of course, water itself. The other half comes from the food you eat. You could live without food for weeks as long as you had water to drink. But you could not live long without water.

Suppose you didn't have running water in your house. You would have to buy containers of water for drinking and cooking. You would have to take your laundry someplace else. You would have to go to a bathhouse to take a bath. You would need a different kind of toilet, one that didn't use water. You would be dirtier. You would spend time and

money solving your problem of getting water.

Modern plumbing delivers running water to sinks, showers, tubs, and toilets. But where does the water come from? How does it get to your house? How can you be sure that it's clean enough to drink? Modern plumbing sends waste water away from your drains and toilets. Where does this water go? How can you be sure that waste water isn't getting mixed up with clean water?

The trip of a drip is very short from your faucet to your drain. It's something you take for granted. But there's a story about this water. It's a story about where it comes from. How it's made clean for use. Where it goes when you've finished with it. It's a story about keeping our water clean and keeping waste water from polluting clean water. And it's a story about money. Pipes and pumps and reservoirs all have to be paid for. Water does not come free.

Want to know the story of the trip of a drip? Read on!

2. Water for a Country Home

A house in the country, far from a city or town, must have its own water supply. Someone living in a cabin in the woods might hand-carry water from a nearby stream. Better yet, the person might pipe in water from a spring. A spring is fresh water that comes from underground. It is usually cleaner than a stream. Spring water is protected from birds and animals that can make a stream unsafe for drinking.

But most country homes get their water from wells. A well is a hole in the ground that reaches water. For hundreds of years people dug wells. If they were lucky, they soon reached wet soil called the *water table*. The water table is the underground wetness that is at the same level as nearby lakes and streams. Rainwater sinks through the ground to the water table. The bottom of a water-table well fills with water from the water table.

Today, most people no longer dig wells to the water table. Instead they *drill* wells into bedrock that lies underneath the water table. They drill until they reach water that is trapped in a water-soaked underground layer of sand or rock called an *aquifer* (ack-wuh-fur). Water from the water table is not as clean as water in the aquifer. Water from rain and melting snow drips down through the soil and the water table into the aquifer. All the seeping and dripping through layers of soil and sand make it very clean. Springs are water that comes to the surface from the aquifer.

An aquifer is like a sink filled with plates and water. Just as water fills in spaces between the plates, in an aquifer, water fills the cracks of rocks and the spaces in gravel and sand. It is an underground basin of wet rocks. It may be surrounded by solid rock or clay that keeps the water from flowing away. The bottom of a well drilled

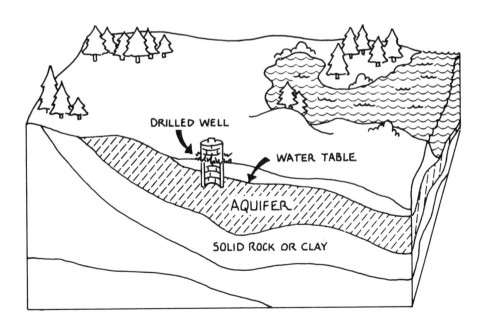

into the aquifer fills with water. Sometimes a drilled well is more than a thousand feet deep.

Once a well has been drilled into the aquifer, how does the water get up to the surface? Years ago, when people used water-table wells, they sent buckets down a well to draw water. They turned a crank, lifting the heavy, water-filled bucket to the surface. Then pumps that raised water were invented. Modern wells have electric pumps that carry water up from the aquifer.

Sometimes the water in the aquifer is trapped and can't flow away. If a well is drilled into it, the pressure on the water makes it spurt out. Sometimes it shoots up in the air. This kind of well is called an *artesian well*. You don't need a pump for an artesian well. There are very few places in the world where you can drill an artesian well.

Suppose you were going to build a house in the country. Where would you drill your well? Your local well-driller knows how deep to drill to reach the aquifer. Where I live, most wells are about 250 feet deep. It doesn't matter much where he drills because he knows that he will strike water.

But suppose you wanted to build a house somewhere on a very large piece of land. The well-driller will look for places that are lower than most of the land. He will look for marshy areas. These places collect rainwater that can seep down to the aquifer.

Some people hire "dowsers" or "water witches" to find a spot for a well. Dowsing is a very strange thing that only a few people are able to do suc-

cessfully. The dowser walks over the land holding a forked branch from a tree. The branch points straight ahead. Suddenly, the branch points toward the ground. The "dowsing rod" sometimes moves so strongly that the dowser's hands are scraped raw. The dowser says that water can be found if a well is drilled where the "dowsing rod" points.

Some people believe in dowsing. Others don't. Science can't explain it. There are lots of stories about dowsers who found water in very dry areas. There are other stories about dowsers who were wrong and pointed out wells that proved to be dry holes. I talked to a well-driller who didn't believe in dowsing. But one of the men who works for him can do it. So if a customer wants a well dowsed, it can be done. Dowsing is a mystery waiting to be solved.

Water from the well is pumped into the house. Cold water goes through pipes directly to the kitchen and bathroom. Some of the water goes to the hot water heater before it is pumped to hot water faucets.

The simplest way to get rid of waste water from a country home is to let it drain into the ground. You just have to make sure that the waste water doesn't drain into the clean water supply. The spot where underground waste water collects is called a *cesspool*. The waste in the cesspool slowly rots. There are millions and millions of tiny plants in the soil that use waste as food. These plants are called *bacteria*. Bacteria can be seen only with a microscope. When bacteria use waste as food, they change the waste into simpler substances that other plants use as fertilizer. Grass that grows over a cesspool is very healthy and green. The land over the cesspool is often soggy. It can also smell bad.

Most modern country homes have *septic tanks*, not cesspools. A septic tank is usually a large metal container that is buried underground. It quietly holds the waste water in one place. The solid waste settles to the bottom. The cleaner water on the top drains into nearby land. A septic tank smells bad only when it is too full and overflows. Every once in a while the septic tank has to be cleaned out. There are companies that can be hired to do this.

People who live in cities and large towns usually don't have their own wells or septic tanks. To find out how they get water and get rid of sewage, keep reading.

SEPTIC TANK SYSTEM

3. Water for Cities and Towns

When experts determine how much water is needed by a city or town, they add up how many people live there. Then they figure that each person needs about 150 gallons of water a day. You need water for more than drinking. You need water for cooking, laundry, bathing, and flushing the toilet. Factories in your area use water. Stores and restaurants use water. Water must be supplied for fighting fires and cleaning streets. Some water ends up in swimming pools. And some is wasted through leaks in underground pipes. All of this water is counted in the 150 gallons for each person.

Cities and towns use billions and billions of gallons every day. It's no small job to deliver clean, drinkable water to all the faucets in a city. Here's how they do it.

Some cities, like Chicago or Montreal, are built on the shores of rivers or large lakes. These lakes and rivers supply their water. Some cities, like Miami and Honolulu, depend on underground water. Still other cities, like New York, Los Angeles, and Denver, pipe in water from higher ground hundreds of miles away. Such water for large cities is stored in lakes called *reservoirs*. Reservoirs, lakes, and rivers are called *surface* water. It's not a good idea to drink water right from a surface water supply. It's usually not clean enough to drink. So before surface water is piped to homes, it is cleaned and treated. This happens at a *water treatment plant.*

Untreated water can have many different kinds of material in it. Each step at the water treatment plant removes a special kind of problem. When water enters the water treatment plant from a reservoir, it first passes through screening. The screen keeps out fish and dead leaves and other large objects that may be floating in the water.

Next the water enters a *flash mixer.* This is a huge tank that has mixing propellers in it. The water is quickly mixed with chemicals. One chemical is *chlorine.* Chlorine is a poisonous yellow gas. It arrives at a treatment plant in heavy metal drums. When it is mixed with water, it quickly dissolves. Chlorine gets rid of certain bad-smelling substances. It also kills bacteria that can make people sick.

Another chemical that is added to the flash mixer is *alum.* Alum helps get rid of very tiny particles that are in the water. These particles are too tiny to settle to the bottom when the water is

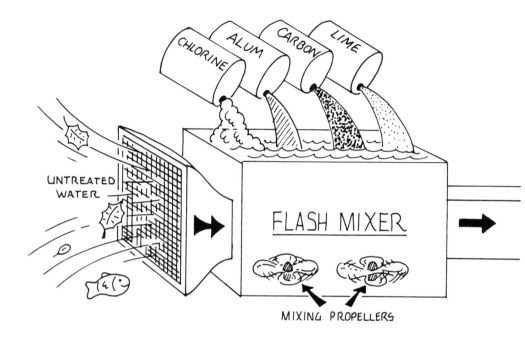

quiet, but they will stick to the alum. Alum that has collected particles is now made up of *flocs*. Flocs are heavy enough to sink to the bottom.

Finely ground up *carbon* may be added. Substances in water that make it smell bad will stick to the carbon particles. Still another chemical, called *lime*, is added to make the water "softer." (I'll tell you more about soft and hard water in a minute.)

After the water is well mixed with chemicals,

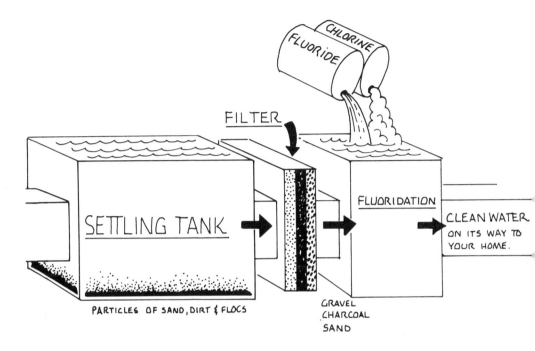

it goes to a huge *settling tank*. Here the water is kept very quiet. All the particles of dirt and sand in the water and the flocs settle to the bottom. The settled material is scraped into a sewer. If you could smell this stuff, it would smell like low tide.

Even at this stage there are substances in the water that are not heavy enough to settle. These particles are removed when the water passes through a *filter*. The filter is made of layers of sand, charcoal, and gravel. However, the water is still not perfectly clean. Very, very tiny particles can pass through the filter along with the water.

Harmful, disease-causing bacteria are among the particles that are small enough to pass through the filter. So before the water treatment plant is finished "making water," more chlorine is added. In many plants, a chemical called *fluoride* is also added at this point. Fluoride has been shown to help keep people from getting cavities in their teeth.

Some water treatment plants spray water into the air. Doing this adds oxygen to the water to make it taste better. Not all plants do this. In many places the water tastes all right without this step.

So, how pure is the water that comes from your faucet? Here's one test you can do. Put some water from your faucet in a saucer. Let the water stand until the dish is dry. (This may take several days.) Carefully look at the dry saucer. Use a magnifying glass if you wish. Run your finger over the surface of the dish. Has any powdery material been left behind? It is made up of *minerals* that are dissolved in the water. Minerals are left behind when the water evaporates into the air.

If you want to be a more exact scientist, do this experiment with two saucers. Put tap water in one saucer and *distilled* water in the other. You can buy a container of distilled water at a hardware store or supermarket. It is used for filling steam irons. The dish with distilled water will have no minerals.

Depending where you live, the water may have more or fewer minerals than water in other parts of the country. The amount of mineral in water makes it "hard" or "soft." "Hard" water got its name because it's "hard" to make suds in it. The minerals in the water combine with soap to make a gray film instead of suds. It is this film that leaves a ring around a bathtub.

Do an experiment to see hard water in action. Put a half cup of cold tap water in each of two jars for which you have lids. Put a half cup of distilled water in a third jar. Put a half teaspoon of Epsom salt in one jar. (You can get Epsom salt at a drugstore.) Put on the lid and shake to mix in the Epsom salt. You have just made some hard water. Now put a half teaspoon of soap flakes in each jar. Make sure that you use soap, not a detergent. (If it is soap, it will say so on the package.) Cover the jars tightly. Give each one five shakes.

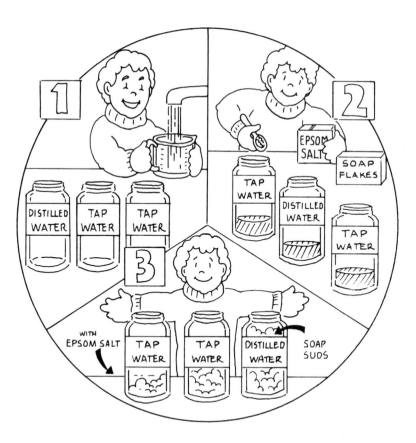

Which one has the most sudsy foam? Which has the least? Distilled water is as soft as you can get. How soft is your tap water?

Minerals in your water will not hurt you. In fact, some mineral water is especially tasty. But there are other substances that can get in the water supply that can cause great illness. *Cholera* is a deadly disease caused by certain bacteria that live in water. People with cholera lose enormous amounts of body fluids because they have severe diarrhea and vomiting. *Typhoid fever* is another deadly disease that can come from germs in an unclean water supply. Enough chlorine is used at water treatment plants to kill these bacteria if they are present.

The people who work at water treatment plants do more than simply hope that their water is safe to drink. They make *sure* their water is safe! Every water treatment plant has a laboratory. A chemist tests the water every single day. No water leaves a water treatment plant without being tested and without passing all tests with flying colors.

The treated water is pumped away from the plant on its way to people's homes. Some of this water goes into hot water heaters. In cities, the water is pumped into large tanks that are often located on tops of buildings. When a faucet is turned on, the falling water flows out strongly. Gravity gives the water enough pressure so that it comes into a sink with some force.

If you live in a city or town, your water is not free. The pipes, the water treatment plants, and the tests that make sure your water is safe all cost money. Workers who help supply you with safe water must be paid. Most water supplies are public. Some of the money comes from taxes paid to the government. Usually this money is used to build or repair pipelines and plants. Public and private water companies also charge people for the water they use. If you live in an apartment, your landlord pays for water. If you live in a private home, your family pays for it. A water meter measures how much water you use so you can be charged the correct amount.

So the next time you turn on your faucet, you will have a good idea of just how far your water has come. But it still has a long way to go.

4. From Sewer to Stream

Raw sewage is the water that leaves drains and toilets. It contains human waste, paper, soap, and detergent. Factories also add waste water to raw sewage. Years ago, the sewers of cities and towns dumped all their raw sewage into rivers and lakes.

They figured that there was enough water in the lakes or rivers to dilute the raw sewage. They were hoping that their raw sewage was like a "drop in a bucket." But they were wrong. Raw sewage *pollutes* water. Polluted water can be poisoned in many ways.

Sometimes the solids in raw sewage settle to the bottom of a lake or river and kill off the plants. Fish can die because they don't have enough to eat. Bacteria in water break down human waste. Some of these bacteria use up the oxygen in the water. Other bacteria, which can live without oxygen, take over. When they use human waste as food, the water becomes dark and smells terrible. Still other bacteria, which come from human waste, find a new home in polluted water. These are the bacteria that cause cholera and typhoid. When there has been a flood, people are told not to drink the water unless they boil it first. That's because the waste water can get mixed with the clean water supply. If the water is boiled, harmful bacteria are killed.

Another ingredient of raw sewage is the *phosphates* in detergents. Phosphates don't directly pollute the water. In fact, they are plant food. Large amounts of phosphates make water plants grow so well that the plants take over a lake or pond. Nothing else can grow. You may have seen such "dead" ponds. They are covered over with a bright green film made of millions of one-celled green plants called *algae* (al-jee).

Factories add to the problem by dumping in chemicals that are poisonous. One such poison is the metal *mercury*. Mercury can show up in the bodies of fish and seafood that people eat. Mercury poisoning comes from eating polluted fish or seafood. Polluted water is not good for drinking, fishing, or swimming. It is a waste. But for many years people polluted water without batting an eyelash, until we finally realized that pollution would catch up to us. It was possible that we would not have enough clean water to drink!

In 1972 and 1974 new laws were passed by the government of the United States. These laws make sure local water companies clean up their sewage so that it can't pollute. They also set the rules for clean, drinkable water. The part of the government that makes sure these laws are obeyed is the Environmental Protection Agency, or EPA. The EPA acts as a watchdog over the safety of the water we drink and the danger of pollution of our waterways.

Since the clean-water laws were passed, many towns and cities have cleaned up their act. They have built *sewage treatment plants*. There's a good chance that the water that leaves your drains and toilet goes into a sewer that takes it to such a sewage treatment plant.

If you went to a modern sewage treatment plant, you would be in for a big surprise. Guess what! It doesn't smell bad! In fact, it doesn't smell at all! If it smells bad, it is not working properly. Here's why.

Most of the solid material in raw sewage is *biodegradable*. This means that it can be broken down by bacteria. Human waste is biodegradable.

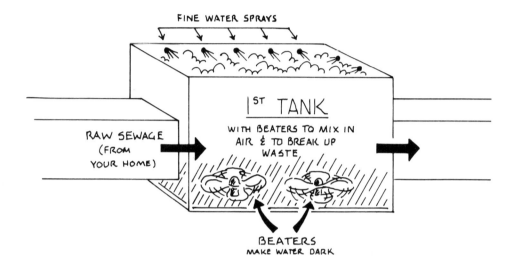

So is paper. So is soap. There are two kinds of bacteria that can feed on raw sewage. One feeds when oxygen is present. These are called *aerobic* bacteria. The other feeds when there is no oxygen present. These are *anaerobic* bacteria, or *anaerobes*. The bad smells are caused by the action of anaerobes.

When the water leaves your drain, there is

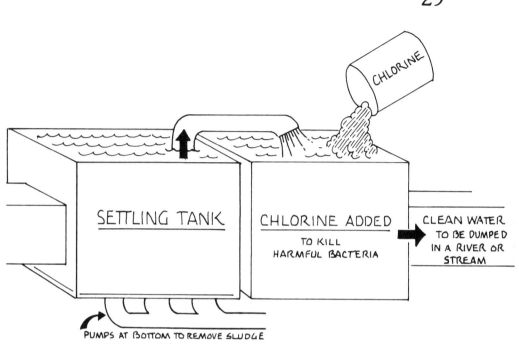

oxygen in it. This means the aerobic bacteria are working. They will keep on working as long as they have oxygen. If they use up all the oxygen, the anaerobes take over. So all a sewage treatment plant must do to stop the anaerobes is to keep adding air to the water.

Air is bubbled into the bottoms of the first tanks raw sewage enters in a sewage treatment plant.

These tanks have many beaters. The beaters constantly stir up the water, mixing the air in. The beating also breaks up the waste into very small pieces that are light enough to spread evenly through the water. The raw sewage tanks look very dirty. The water is brown, the color of human solid waste. There is foam on the surface from all the soap and detergents. Jets around the edges of the tanks send out a fine spray of clean water all along the surface. This keeps the foam from climbing up the sides of the tank.

All the oxygen in the water and the waste material make the aerobic bacteria grow. Millions and millions of bacteria are attracted to the tiny pieces of waste in the water. They stick to the pieces of waste, which then become heavy. Perfect... because the sewage moves on to the next tank, a *settling* tank, where it will remain quiet for several hours. The waste particles, loaded with bacteria, are *flocs*. Flocs, remember, are particles that are heavy enough to settle to the bottom. Alum makes flocs in water treatment. Aerobic bacteria make flocs in sewage treatment.

Water is pumped off the top of the settling tanks. It is fairly clean. Some sewage treatment plants now filter this water. Chlorine is added to kill harmful bacteria. The water that finally comes out of a sewage treatment plant is often so clean that it can be dumped into a nearby river or stream. The plant I visited produced water safe enough to drink. It emptied into a nearby trout stream.

The material that settles to the bottom of a settling tank is called *sludge*. Another job of sewage treatment plants is to collect sludge. Now the problem is to get rid of the sludge. The first part is easy. The sludge is quite liquid. Pumps remove it from the bottom of tanks. The sludge is made thicker by putting it in a separate settling tank. After a day, a large amount of cleaner water separates from the sludge. It rises to the top. This water is pumped back into the sewage flow.

In some places, the thickened sludge is packed

into plastic bags and taken to landfill dumps, where it will be buried. But a big city, like New York, doesn't dump its sludge in land. Instead, the sludge is allowed to be digested by anaerobes. About half of the sludge is broken down into water and gases. One of these gases is *methane,* which is a fuel. Some sewage treatment plants burn their methane to make enough electricity to run their plant. The digested sludge is loaded on boats and dumped

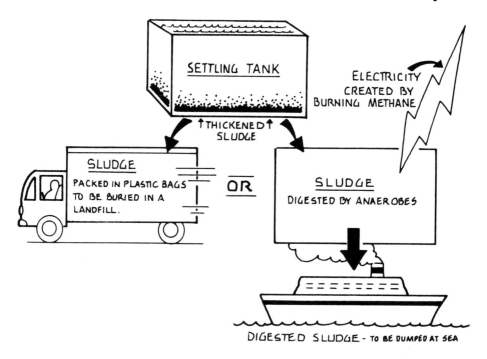

in the sea, miles away from the shore, where we hope that it won't cause pollution problems.

Sewage treatment is simply this: you have a mixture of water and solids. Treatment separates the mixture into water and sludge. The water is now free to continue its trip. No one has to pay for the next part of its journey.

5.
Traveling in Circles

A drop of water that was once in your house can go many places after leaving a sewage treatment plant. It can travel from one stream to another. It can travel through lakes and rivers. It can become part of the ocean. It can also turn up again in your faucet. Water travels in circles. Nature is the greatest recycler of all time. All because water is a truly amazing substance.

Water is a pure substance made of *molecules*. A water molecule is the smallest particle of water that is still water. A molecule is so small that no one has ever seen one, even with the strongest microscope. Scientists have other ways of measuring the size of molecules. They also know how water molecules act.

The water that comes out of your faucet is a liquid. If you pour some water into a container, it will take the *shape* of the container. But it will not

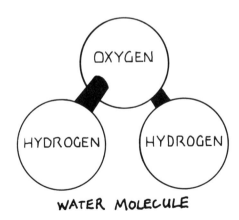
WATER MOLECULE

spread out and take the *size* of the container and fill it completely. Liquid water has a definite surface. Liquid water molecules are always rolling over each other and bumping into each other.

If you put some liquid water in your freezer, the molecules begin acting differently. They slow down. When the water temperature reaches 32 degrees Fahrenheit, the molecules stop tumbling over each other. They are locked in place. Water is no longer a liquid. It is a clear, colorless solid you know as ice. Solid ice has its own size and shape no matter what kind of container you put it in.

When you heat liquid water, the opposite thing happens. The molecules begin moving faster and faster. They move away from each other. Bubbles form on the bottom of the pan. The bubbles rise to the surface. The water boils. The molecules escape into the air. They have become a gas you call *steam*. A gas is somewhat like a liquid. It takes the shape of its container, but it also takes its size. Suppose the steam doesn't go into a container. Suppose it goes into the air. The water molecules just keep on traveling. They become one of the gases of the air. Only now they are known as *water vapor*, not steam.

You don't have to boil water to change it into a gas. Water molecules are always escaping from the surfaces of liquid water into the air. This change is called *evaporation*. Water is always evaporating from the surfaces of rivers and lakes and the oceans. Water that once passed through your house could wind up becoming water vapor. Water vapor in the air collects as water droplets in clouds. Heavier and heavier droplets can collect in clouds. When the drops are heavy enough, they return to earth as rain or sleet or snow. Water from clouds can wind up in a reservoir or aquifer and find its way back to your house.

Actually, it's not likely that any particular water molecule makes a round trip through your house. There are too many water molecules and the world is too big a place. People use a very small amount

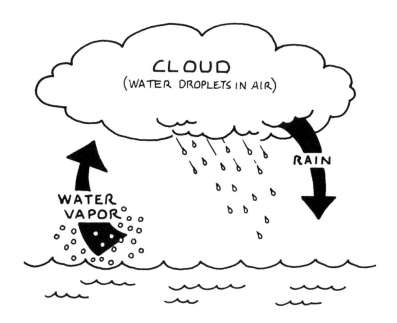

of the world's fresh-water supply. For every hundred drops of rain, only two drops are likely to come out of a faucet. The rest of the water soaks into the ground, waters crops, or runs off in rivers and lakes.

The trip of a drip that comes to your faucet is a long one. What about the last, tiny leg of its journey from faucet to drain? Maybe you have a dripping faucet in your house. A dripping faucet can waste a lot of water, and someone is paying for it. Now you know how important it is to have it fixed. But before you do, go take a careful look at a drip. If your faucet doesn't drip, turn it on a tiny amount so that a slow drip develops.

Here's what you see. Water collects on the tip of the faucet in a hanging bag. The bag gets longer and longer until it is heavy enough to break away. Then it falls to the drain in a tear-shaped drop. The water molecules on the surface of the drop pull together. They act like a skin. If your drip fell on the moon, where there is no air, it would not have the shape of a tear. It would be a perfect little ball. The motion of the air around a drop of water changes the ball into a teardrop shape.

The water in your house is a world traveler. You have followed all the steps of its journey. Now you know its story. Yea!

Some Tricks with Your Drips

Want to fool your friends and amaze your parents and teachers? Water *is* awesome stuff, and you can prove it. All you need is some inside tips on your drips. And they're coming right up. You're going to find out how science can help you do things that seem impossible. Follow the directions and take advantage of some things that science has discovered about water. No one will believe you when you tell them what you can do. But it's a "Gotcha!" every time, I promise.

BET YOUR FRIEND CAN'T MAKE STYROFOAM FLOAT IN THE CENTER OF A GLASS OF WATER. BUT YOU CAN!

All you need is a nearly full glass of water and

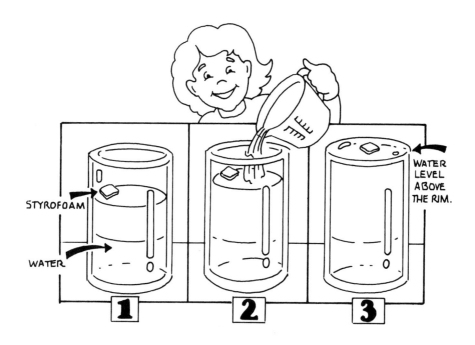

a piece of Styrofoam about the size of your little fingernail. You can cut it out of a Styrofoam cup.

Drop the Styrofoam into the water. It will float at the edge of the surface, next to the glass. Dare your friend to make it stay in the center. He or she may push it to the middle. The minute the finger is removed, bingo! The Styrofoam moves "home" to the edge.

Now you, the wise one, make it happen. Carefully add more water to the glass. You might want to use a measuring cup with a spout to do this. Keep adding until the level of water goes above the rim of the glass. Be careful not to shake the table, or you'll spill it. The water will bulge at the surface. The Styrofoam moves to the center. You win!

Here's what's happening. Remember that water has a "skin" where it meets the air. This skin is called *surface tension*. Surface tension is caused by the strong attraction water molecules have for each other. They pull together at the surface. It's as if they are trying to shrink the surface and make it as small as possible.

When the glass isn't quite full, surface tension isn't the only force of attraction at work. Look carefully at the place where the water meets the glass. Instead of bulging, the water curves up so that the water closest to the glass is higher than the rest of the surface. This happens because water molecules are more attracted to the glass than they are to each other.

This attraction between water and glass makes it possible for the water to wet the glass. So when the glass is not full, all the surface water molecules are pulling toward the edge. The piece of Styrofoam is caught up in this pull and stays next to the glass where the pull is strongest.

When you overfill the glass, surface tension pulls toward the center. Again the Styrofoam gets caught by the pull. Its new home is the center. It's here that the pull is now strongest. Naturally, you now can't get it to stay by the edge.

45

You can see the motion of molecules! You can see that some molecules move faster than others.

You can't see molecules. There are billions in each drop of water. So, how can you see their motion? The answer is that you use something you can see, like a dye. If the dye moves, you

COLD WATER   ROOM TEMPERATURE WATER   HOT WATER

know it does so because it is being pushed around by collisions with countless numbers of water molecules.

Here's how to set up your experiment. You will need three glasses and a package of powdered grape drink. Put ice-cold water in one glass. Use an ice cube, if you wish. Put water that is at room temperature in a second glass. Put very hot tap water in the third glass. Let all three glasses rest on a table so that the water seems quiet. Now drop a pinch of powdered grape drink into each glass. Watch.

The purple dye travels through each glass of water. Where does it move the fastest? Where does it move most slowly? What does this tell you about temperature and the speed of molecules? When you boil water, why does it move from the bottom of a pan to the top? (Hint: think about where the heat is.)

Let the glasses stand. Sooner or later the purple dye spreads evenly throughout each glass. Moving water molecules spread out grape drink molecules. This spreading of molecules is called *diffusion.* In this experiment you see the diffusion in liquids. There is also diffusion in gases. Every smell your nose detects comes to you by diffusion. The molecules you smell diffuse through the air.

BET YOU CAN MOVE WATER BACK AND FORTH BETWEEN TWO GLASSES WITHOUT POURING IT!

All you need to do this trick is about two feet of tubing. You can buy tubing at a store that supplies aquariums. Fill one glass with water. Put it on a table. Put the other, empty glass on a chair seat near the table. Put one end of the tubing in the glass of water and hold it there. Suck the other end of the tubing. When the water reaches your mouth, quickly put the end of the tube in the second glass. Water will run up through the tube and then down to the second glass.

This kind of tube is called a *siphon.* When you have siphoned off about half of the water in the glass on the table, raise the glass on the chair. Be careful to keep both ends of the siphon under-

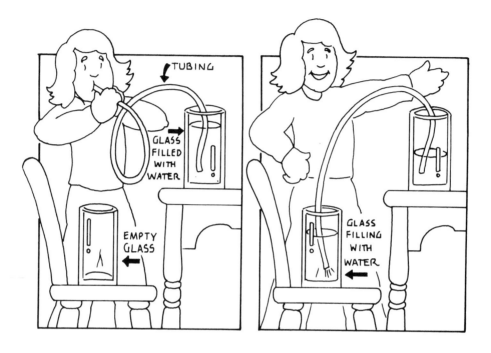

water. What happens when the glass on the chair is above the glass on the table? What happens when both glasses are at the same level? Can you figure out how to get the water to flow back and forth between glasses?

Here's why a siphon works. First of all, water flows from higher places to lower places. It's no different from any other kind of substance on earth. Gravity acts on them all. That's why most of the rivers on earth flow toward the sea. A siphon works only when the tube is completely full of water. Then the water molecules stick together and go with the flow. They stick together enough to travel upward a little way, as long as the main pull is down. A siphon does not work if there is any air in the tube. Break the connection between water molecules with an air bubble, and your siphon doesn't work.

BET YOU CAN MELT AN ICE CUBE IN AN OVERFULL GLASS WITHOUT DRIPPING A DROP!

Pour water in a glass until it is almost full. Put a large ice cube in the glass. It will float. Add water until the surface is bulging above the rim of the glass. The top of the ice cube is above the surface. Let it melt. Not a drop will spill over the rim of the glass. Not only that! The water level will not change at all.

Here's why. Melted ice takes up less space as water than it took up as ice. A floating ice cube is partly underwater. The underwater part of an ice

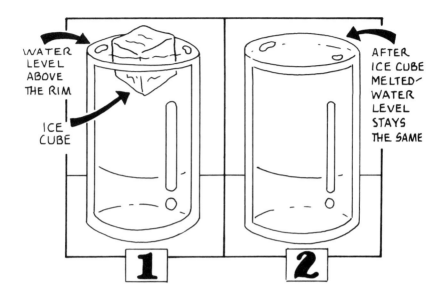

cube takes up a certain amount of space. When the ice cube melts, it shrinks to exactly this same space. When water freezes, it takes up more space. That's why you don't want to freeze a sealed bottle of liquid. Expanding ice can crack the bottle.

It's a good thing ice floats. If it didn't, the ice at the North and South poles would be under water. The oceans would overflow. All the land on earth would be at the bottom of the sea!